Cat Dissection
A Laboratory Guide

Second Edition

CONNIE ALLEN

VALERIE HARPER

Edison College

John Wiley & Sons, Inc.

Cat Dissection
A Laboratory Guide

Second Edition

CONNIE ALLEN

VALERIE HARPER

Edison College

BICENTENNIAL
BICENTENNIAL
1807
WILEY
2007
BICENTENNIAL
BICENTENNIAL

John Wiley & Sons, Inc.

SENIOR EXECUTIVE EDITOR	Bonnie Roesch
PROJECT EDITOR	Mary O'Sullivan
EXECUTIVE MARKETING MANAGER	Clay Stone
PRODUCTION MANAGER	Pamela Kennedy
PRODUCTION EDITOR	Sarah Wolfman-Robichaud
SENIOR DESIGNER	Kevin Murphy
SENIOR ILLUSTRATION EDITOR	Anna Melhorn
PHOTO EDITOR	Felicia Ruocco
SENIOR MEDIA EDITOR	Linda Muriello
MEDIA PROJECT MANAGER	Steve Chasey
COVER DESIGN	Carol Grobe
INTERIOR DESIGN	Nancy Field
COPYEDITOR	Kathy Drasky
PROJECT MANAGER	Jane Shifflet/GTS Companies: PA/York Campus

This book was typeset in 10/12 Times Roman by GTS Companies and printed and bound by Von Hoffmann Press. The cover was printed by Von Hoffmann Press.

The paper in this book was manufactured by a mill whose forest management programs include sustained yield harvesting of its timberlands. Sustained yield harvesting principles ensure that the number of trees cut each year does not exceed the amount of new growth.

The procedures in this text are intended for use only by students with appropriate faculty supervision. In preparing the text, care has been taken to identify potentially hazardous steps and to insert safety precautions where appropriate. The authors and publisher believe the procedures to be useful tools if performed with the materials and equipment specified, in careful accordance with the instructions and methods in the text. However, these procedures must be conducted at one's own risk. The author and publisher do not warrant or guarantee the safety of individuals using these procedures and specifically disclaim any and all liability resulting directly or indirectly from the use or application of any information contained in this book.

This book is printed on acid free paper. ∞

ISBN 978-0-471-70141-5, 0-471-70141-6

Printed in the United States of America

10 9 8

Cat Dissection A Laboratory Guide, 2nd ed.

Outline

Preface

A. Preparing the Cat

1. With gloves on, remove the cat from its bag and lay the cat on a dissecting tray. Keep any liquid preserving solution that remains in the bag.

2. Review the directional terms for the cat in Figure CP.1. Note the differences between four-legged animals and humans.
   - *Anterior* is toward the cephalic (head) end of the cat.
   - *Posterior* is toward the caudal (tail) end of the cat.
   - *Superior* is toward the dorsal (back) surface.
   - *Inferior* is toward the ventral (belly) surface.

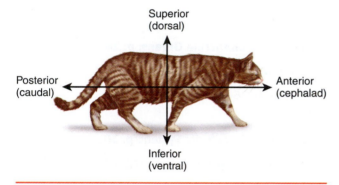

Superior
(dorsal)

Posterior
(caudal)

Anterior
(cephalad)

Inferior
(ventral)

FIGURE CP.1   Directional terminology for the cat.

3. Place your cat ventral surface up on the dissecting tray.

4. Identify the gender of your cat. Males have a scrotum and a prepuce, a small mound anterior to the scrotum in which the penis is located. Females have a urogenital aperture, an opening located anterior to the anus that is a common passageway for the urinary and reproductive systems. Four or five teats (nipples) are present on both male and female cats. Be able to identify both sexes externally.

5. Prepare a label for your cat with the names of your group members and the gender of your cat.

6. Follow the instructions for skinning the cat (Figure CP.2a) if you are dissecting skeletal muscles, or the instructions for opening the ventral body cavities (Figure CP.2b) if you are dissecting an organ system.

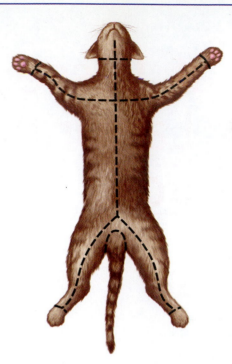

(a) Incisions for skinning

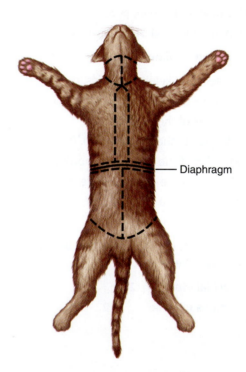

Diaphragm

(b) Opening ventral body cavities

----- Incision line

FIGURE CP.2   Cat incisions.

B. Removing the Skin

1. Referring to Figure CP.2a, pinch the skin on the ventral surface of the neck. Using scissors, carefully make a small, longitudinal incision at the midline through the skin only. Use care not to cut into the underlying muscle layer.

2. Continue cutting longitudinally along the midline toward the lower lip and then posteriorly, stopping anterior to the genital area.

3. Cut the skin around the neck.

4. Make a horizontal cut across the chest and continue cutting down the midline of the extremities as indicated in Figure CP.2a. Make diagonal cuts in the groin and continue midline down the extremities. Cut the skin around all paws.

5. Use your fingers to carefully peel the skin from the underlying muscles. Cutaneous muscles, such as the platysma, are attached to the undersurface of the skin and will be removed as you peel away the skin.

6. Continue peeling the skin until it is only attached at the face and the tail. Cut around the base of the tail, leaving the skin on the tail. Cut the skin around the face of the cat, leaving the skin on the face, ears, and forehead. Peel the skin from the head and save it.

7. Carefully remove as much fat and superficial fascia as possible with your fingers or forceps.

8. Wrap the skin around the cat and follow your instructor's directions for storing your cat in the plastic bag. The skin will prevent the tissues from drying out and prevent the growth of bacteria and mold. Dispose of fascia and fat as indicated by your instructor. Do not forget to attach the label identifying your cat before storing it.

C. Opening Ventral Body Cavities

1. At the midline, just above the pubic bone, carefully make a longitudinal incision through the abdominal muscles. Refer to Figure CP.2b. Continue the incision to the ribs.

2. Cut either to the right or left of the sternum, cutting through the costal cartilages. Continue cutting midline through the neck.

3. Cut horizontal incisions at the top and at the base of the neck.

4. Cut horizontal incisions anterior and posterior to the diaphragm as indicated in Figure CP.2b, and cut the diaphragm away from the ventral body wall. Open the flaps to expose the thoracic and abdominal cavities, leaving the diaphragm intact.

5. Use a scalpel to make a longitudinal cut down each inner wall of the rib cage. Bend the walls outward to break the ribs, allowing the flaps of the thoracic wall to stay open.

6. Dispose of fascia and fat as indicated by your instructor. Do not forget to attach the label identifying your cat before storing it.

Dissection 1: Skeletal Muscles

Many skeletal muscles of the cat are similar to human muscles. This dissection will reinforce your knowledge of human skeletal muscles and allow you to observe the fascia that surrounds, protects, and compartmentalizes these muscles. Assemble your dissection equipment and safety glasses, put on your gloves, and obtain your cat. Position your cat within the dissection tray, including the tail. Keep any remaining preserving fluid in the bag to keep your cat moist and inhibit bacterial and mold growth.

Procedure

A. Dissecting Skeletal Muscles

It is important to carefully remove the fascia to observe the individual muscles. However, using scissors or scalpels may result in cutting muscles or other structures. Blunt dis-

section is a technique that uses blunt probes and forceps to remove fascia and separate muscles. To observe a deep muscle, you will have to cut the superficial muscle at the midline and reflect (pull back) the edges toward the origin and insertion.

B. Muscles of the Head and Neck

1. Refer to Figure C1.1 to locate the following superficial muscles on the cat. Cats have a platysma, but this muscle was most probably removed during the skinning process.
   - Masseter
   - Digastric
   - Mylohyoid
   - Sternohyoid
   - Sternothyroid
   - Sternomastoid (sternocleidomastoid in humans)

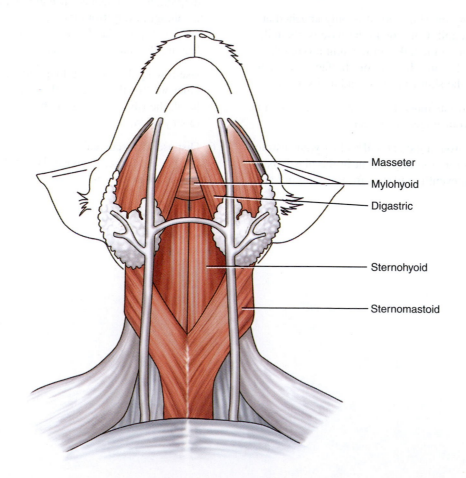

Masseter
Mylohyoid
Digastric

Sternohyoid

Sternomastoid

FIGURE C1.1 Superficial muscles of the head and neck.

Ventral

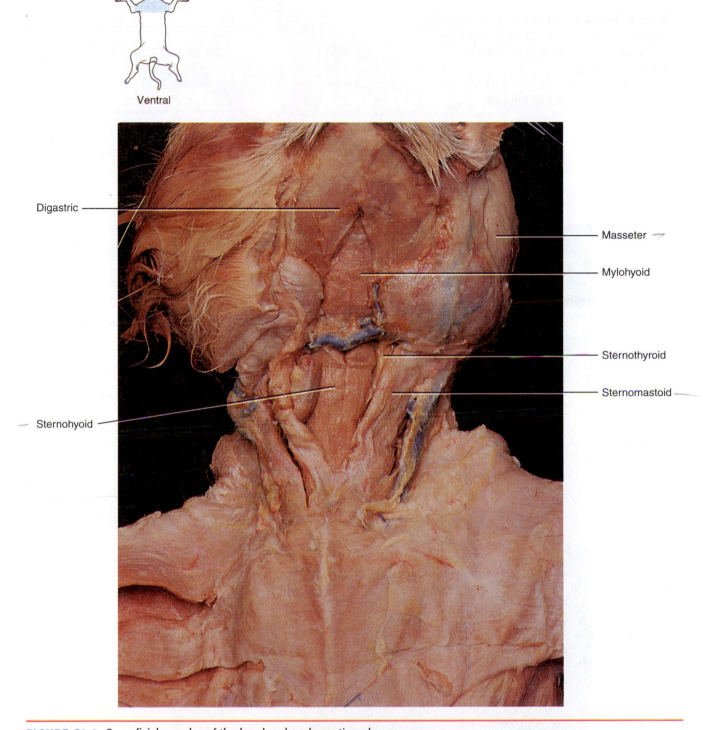

Digastric

Masseter

Mylohyoid

Sternothyroid

Sternomastoid

Sternohyoid

FIGURE C1.1 Superficial muscles of the head and neck, *continued*.

C. Muscles of the Chest

1. Refer to Figure C1.2a to locate the following
   superficial muscles on the chest of the cat:
   • Pectoantebrachialis (not in humans)
   • Pectoralis major
   • Pectoralis minor
   • Xiphihumeralis (not in humans)

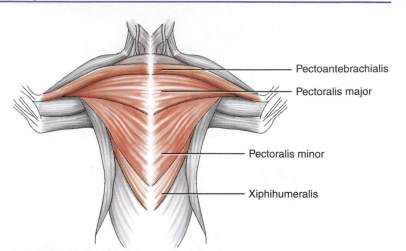

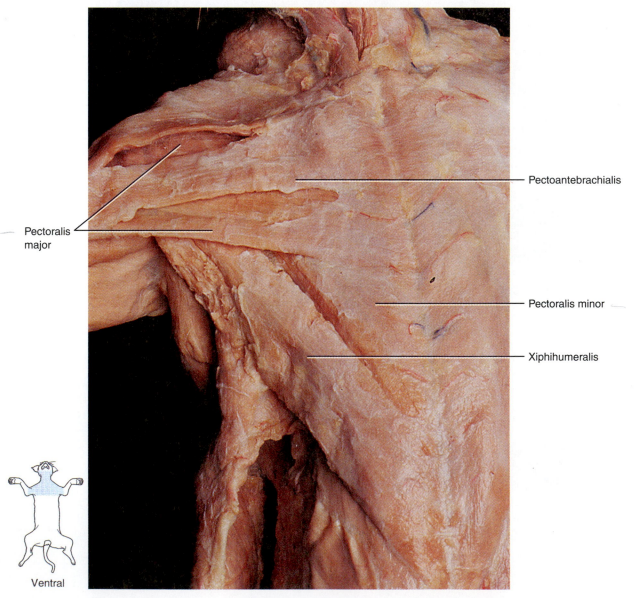

Ventral

(a) Superficial muscles

FIGURE C1.2   Muscles of the chest.

2. Cut and reflect the pectoralis major, pectoralis minor, and the xiphihumeralis.

3. Refer to Figure C1.2b to locate the following deep muscles on the ventral thorax of the cat:
   • External intercostals
   • Serratus ventralis (serratus anterior in humans)

4. If advised by your instructor, cut and reflect these muscles to observe the internal intercostal muscles that run obliquely to the external intercostals.

Ventral

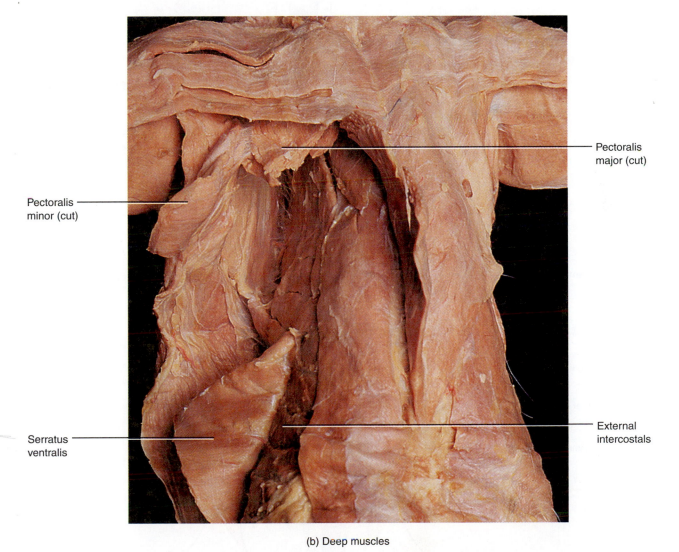

Pectoralis major (cut)

Pectoralis minor (cut)

Serratus ventralis

External intercostals

(b) Deep muscles

FIGURE C1.2 Muscles of the chest, *continued*.

D. Muscles of the Abdomen

1. Refer to Figure C1.3 to locate the following superficial muscles on the abdomen of the cat:
   • Rectus abdominis
   • External oblique

2. Cut and reflect the very thin external oblique to observe the underlying:
   • Internal oblique

3. Cut and reflect the very thin internal oblique to observe the underlying:
   • Transverse abdominis; often, the transverse abdominis is attached to the underside of the internal oblique.

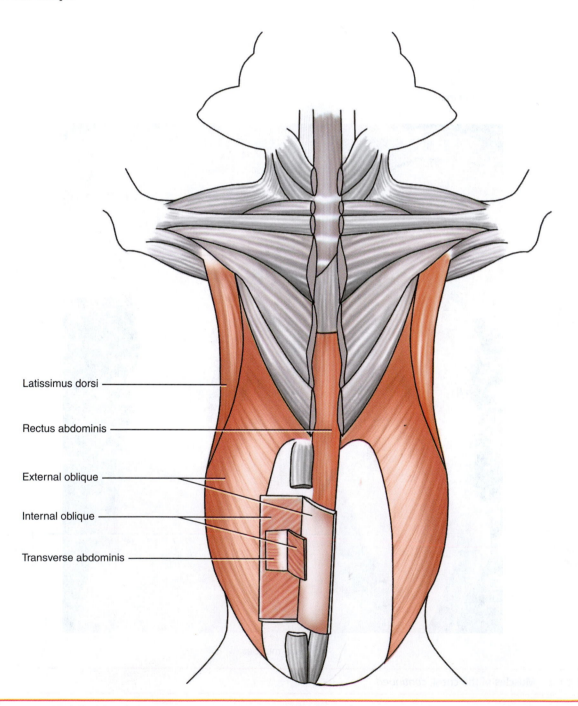

Latissimus dorsi

Rectus abdominis

External oblique

Internal oblique

Transverse abdominis

FIGURE C1.3   Muscles of the abdomen.

Ventral

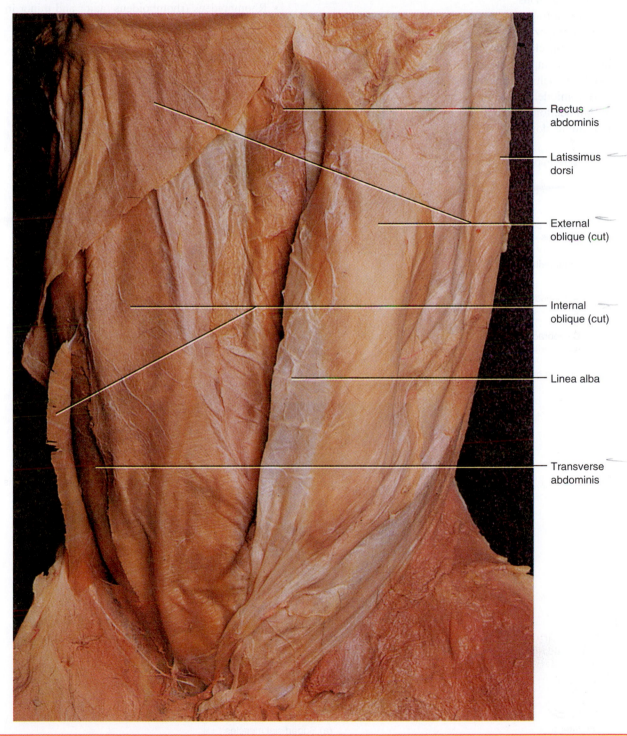

Rectus abdominis

Latissimus dorsi

External oblique (cut)

Internal oblique (cut)

Linea alba

Transverse abdominis

FIGURE C1.3   Muscles of the abdomen, *continued*.

E. Muscles of the Back and Shoulder

1. Refer to Figure C1.4a to locate the following superficial muscles:
   - Trapezius muscles—The cat has three separate muscles, compared with a single human trapezius.
     —Clavotrapezius
     —Acromiotrapezius
     —Spinotrapezius
   - Deltoid muscles—The cat has three separate deltoid muscles, compared with one in humans.
     —Clavobrachialis (clavodeltoid)
     —Acromiodeltoid
     —Spinodeltoid
   - Latissimus dorsi

2. Cut and reflect the trapezius muscles and the latissimus dorsi.

3. Refer to Figure C1.4b to locate the following deep muscles:
   - Splenius
   - Levator scapulae ventralis (levator scapulae in humans)
   - Rhomboideus capitis (not in humans)
   - Rhomboideus (rhomboideus major and minor in humans)
   - Supraspinatus
   - Infraspinatus
   - Teres major

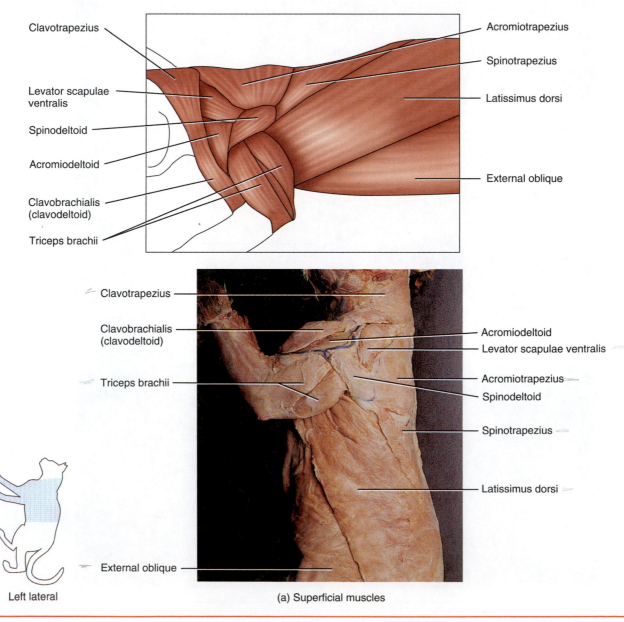

FIGURE C1.4   Muscles of the shoulder.

Left lateral

(a) Superficial muscles

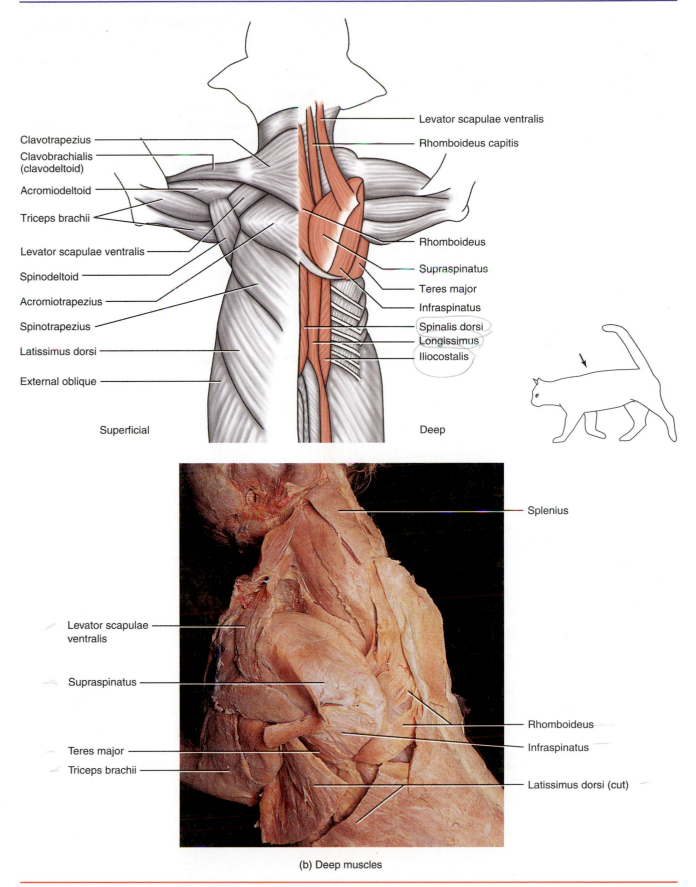

Levator scapulae ventralis

Rhomboideus capitis

Clavotrapezius

Clavobrachialis
(clavodeltoid)

Acromiodeltoid

Triceps brachii

Levator scapulae ventralis

Spinodeltoid

Acromiotrapezius

Spinotrapezius

Latissimus dorsi

External oblique

Rhomboideus

Supraspinatus

Teres major

Infraspinatus

Spinalis dorsi

Longissimus

Iliocostalis

Superficial

Deep

Splenius

Levator scapulae
ventralis

Supraspinatus

Rhomboideus

Infraspinatus

Teres major

Triceps brachii

Latissimus dorsi (cut)

(b) Deep muscles

FIGURE C1.4   Muscles of the shoulder, *continued*.

F. Muscles of the Arm and Forearm

1. Using Figure C1.5a, locate the following muscles on the lateral arm:
   - Brachialis
   - Triceps brachii lateral head
   - Triceps brachii long head

2. Cut and reflect the lateral head of the triceps brachii muscle and identify the:
   - Triceps brachii medial head

3. Using Figure C1.5a, locate the following muscles on the lateral forearm. These muscles are listed from anterior to posterior:
   - Brachioradialis
   - Extensor carpi radialis longus
   - Extensor digitorum communis
   - Extensor digitorum lateralis
   - Extensor carpi ulnaris

4. Lift the extensor carpi radialis longus to observe the underlying muscle (see Figure C1.5b):
   - Extensor carpi radialis brevis

5. Using Figure C1.5b, locate the following muscles on the medial arm:
   - Biceps brachii—Cut and reflect the pectoante-brachialis muscle to better observe the biceps brachii
   - Epitrochlearis (not in humans)

6. Using Figure C1.5b, locate the following muscles on the medial forearm:
   - Flexor carpi radialis
   - Palmaris longus
   - Flexor carpi ulnaris
   - Pronator teres

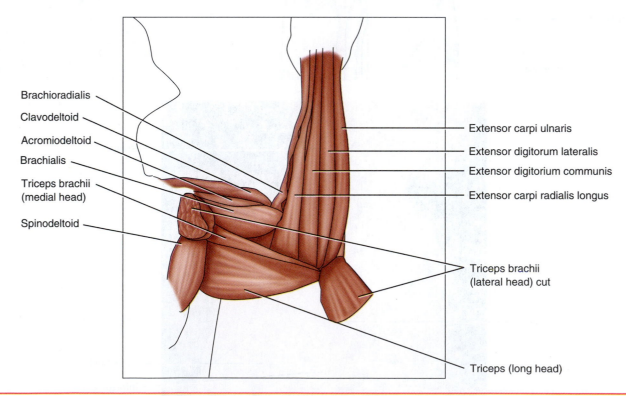

FIGURE C1.5 Muscles of the arm and forearm.

Dorsal

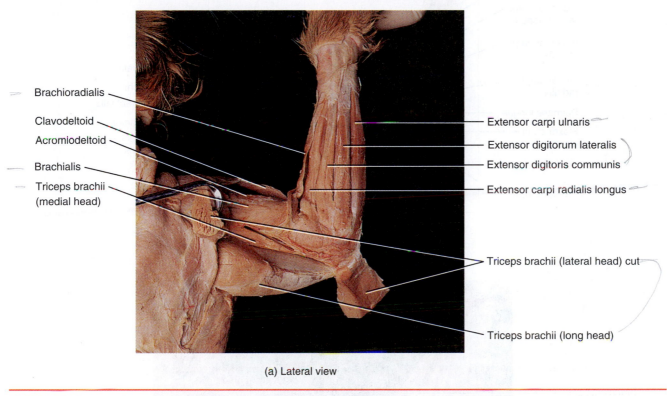

Brachioradialis

Clavodeltoid

Acromiodeltoid

Brachialis

Triceps brachii
(medial head)

Extensor carpi ulnaris

Extensor digitorum lateralis

Extensor digitoris communis

Extensor carpi radialis longus

Triceps brachii (lateral head) cut

Triceps brachii (long head)

(a) Lateral view

FIGURE C1.5   Muscles of the arm and forearm, *continued*.

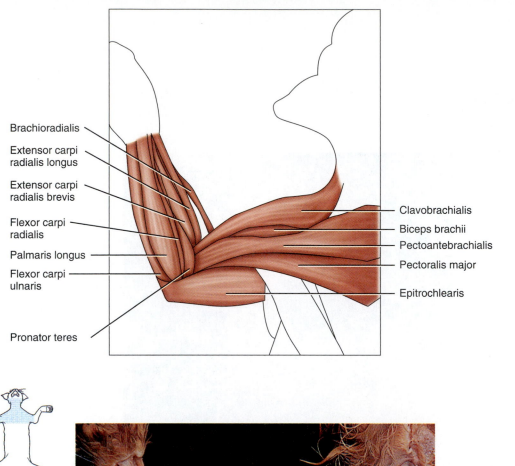

Brachioradialis
Extensor carpi radialis longus
Extensor carpi radialis brevis
Flexor carpi radialis
Palmaris longus
Flexor carpi ulnaris
Pronator teres

Clavobrachialis
Biceps brachii
Pectoantebrachialis
Pectoralis major
Epitrochlearis

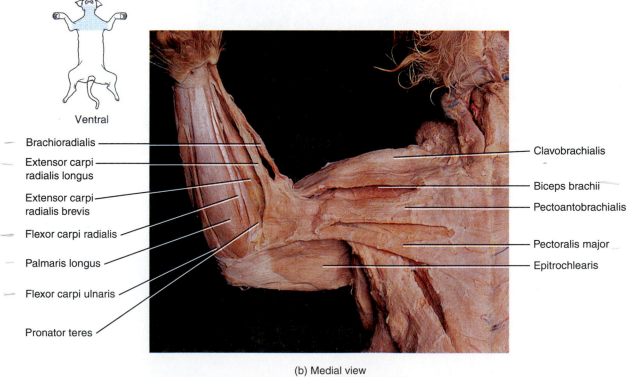

Ventral

Brachioradialis
Extensor carpi radialis longus
Extensor carpi radialis brevis
Flexor carpi radialis
Palmaris longus
Flexor carpi ulnaris
Pronator teres

Clavobrachialis
Biceps brachii
Pectoantobrachialis
Pectoralis major
Epitrochlearis

(b) Medial view

FIGURE C1.5   Muscles of the arm and forearm, *continued.*

G. Muscles of the Thigh

1. Thighs of four-legged animals have broad lateral and medial surfaces. Note how the quadriceps and hamstring muscles are distributed on the lateral and medial surfaces of the cat, and compare this with the distribution in humans. Using Figure C1.6a, locate the following superficial muscles on the lateral thigh:
   - Sartorius
   - Tensor fasciae latae
   - Gluteus medius
   - Gluteus maximus
   - Caudofemoralis (not in humans)
   - Vastus lateralis
   - Biceps femoris
   - Semitendinosus

2. Using Figure C1.6b, locate the following superficial muscles on the medial thigh:
   - Sartorius
   - Adductors
   - Gracilis

3. Cut and reflect the sartorius and the gracilis muscles.

4. Using Figure C1.6c, locate the following deep muscles on the medial thigh:
   - Iliopsoas
   - Pectineus
   - Adductor longus
   - Adductor femoris (adductor magnus in humans)
   - Vastus lateralis
   - Rectus femoris
   - Vastus medialis
   - Semimembranosous

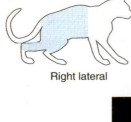

Right lateral

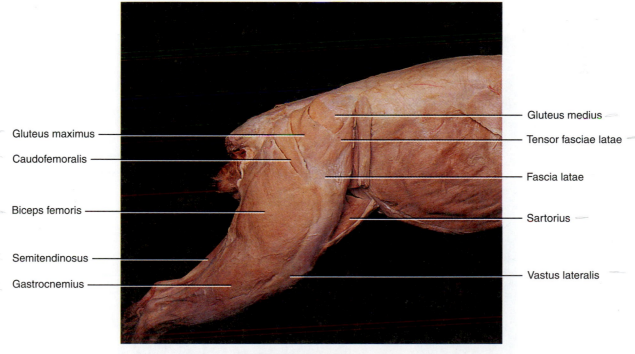

(a) Superficial muscles, lateral view

FIGURE C1.6   Muscles of the thigh.

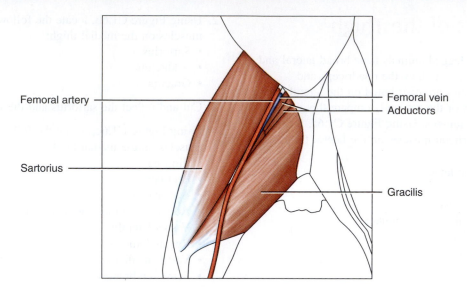

Femoral artery

Sartorius

Femoral vein
Adductors

Gracilis

Ventral

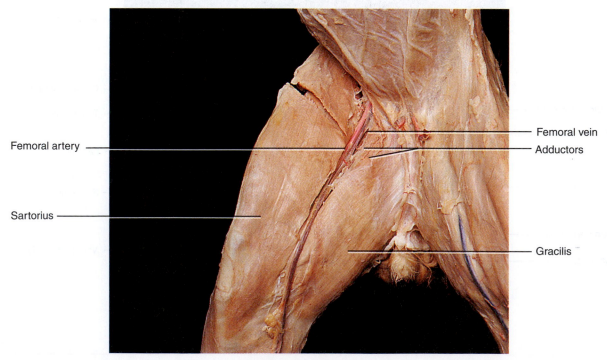

Femoral artery

Sartorius

Femoral vein
Adductors

Gracilis

(b) Superficial muscles, medial view

FIGURE C1.6   Muscles of the thigh, *continued*.

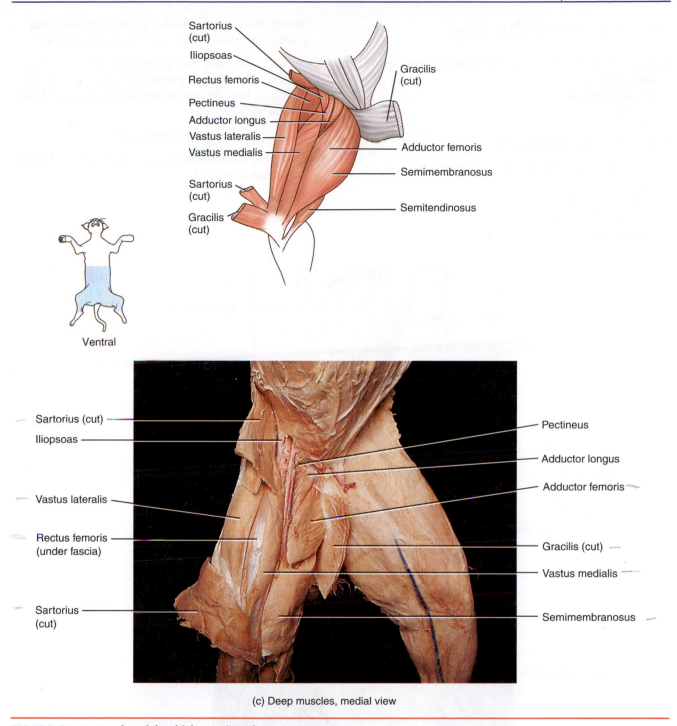

Sartorius (cut)
Iliopsoas
Rectus femoris
Pectineus
Adductor longus
Vastus lateralis
Vastus medialis
Sartorius (cut)
Gracilis (cut)

Gracilis (cut)
Adductor femoris
Semimembranosus
Semitendinosus

Ventral

Sartorius (cut)
Iliopsoas
Vastus lateralis
Rectus femoris (under fascia)
Sartorius (cut)

Pectineus
Adductor longus
Adductor femoris
Gracilis (cut)
Vastus medialis
Semimembranosus

(c) Deep muscles, medial view

FIGURE C1.6   Muscles of the thigh, *continued*.

H. Muscles of the Leg

1. Using Figure C1.7a, locate the following muscles on the lateral leg:
   • Gastrocnemius
   • Soleus
   • Peroneus
   • Extensor digitorum longus

2. Using Figure C1.7b, locate the following muscles on the medial leg:
   • Tibialis anterior
   • Flexor digitorum
   • Gastrocnemius

3. Identify the calcaneal tendon (Achilles tendon) that attaches the gastrocnemius to the calcaneal bone.

4. Place the skin back over your cat and follow your instructor's directions to prepare the cat for storage in the plastic bag. Be sure to attach your group's identification tag.

5. Clean your tabletop with disinfectant.

6. Wash your dissection tools, dissection tray, and hands before leaving the lab.

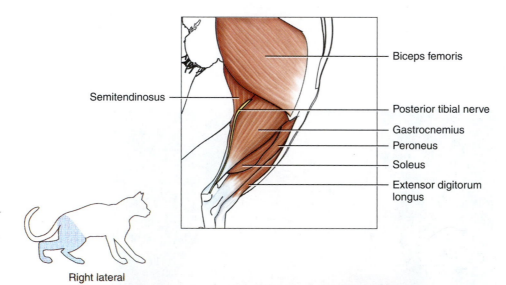

Biceps femoris
Semitendinosus
Posterior tibial nerve
Gastrocnemius
Peroneus
Soleus
Extensor digitorum longus

Right lateral

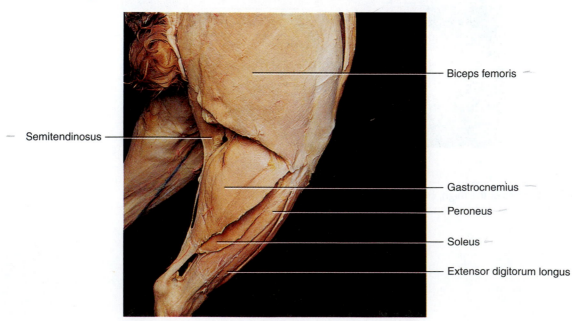

Biceps femoris
Semitendinosus
Gastrocnemius
Peroneus
Soleus
Extensor digitorum longus

(a) Lateral view

FIGURE C1.7 Muscles of the leg.

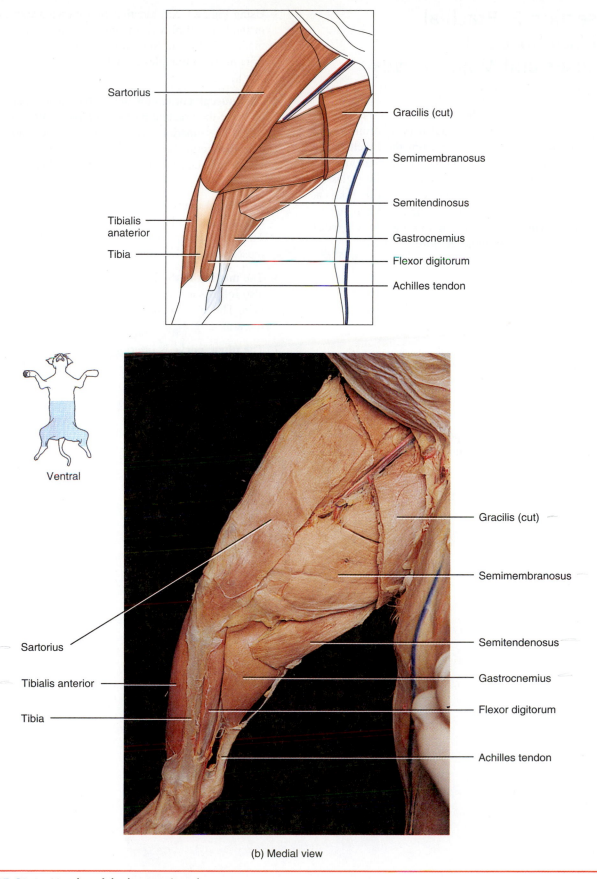

Sartorius

Gracilis (cut)

Semimembranosus

Semitendinosus

Tibialis anaterior

Gastrocnemius

Tibia

Flexor digitorum

Achilles tendon

Ventral

Gracilis (cut)

Semimembranosus

Semitendenosus

Sartorius

Tibialis anterior

Gastrocnemius

Tibia

Flexor digitorum

Achilles tendon

(b) Medial view

FIGURE C1.7   Muscles of the leg, *continued*.

Dissection 2: Brachial and Lumbosacral Plexuses and Major Nerves

This dissection illustrates the structure of a plexus. You will observe the network of spinal nerves forming each plexus. The major nerves arising from the brachial and lumbosacral plexuses are the same as in the human.

Assemble your dissection equipment and safety glasses, put on your gloves, and obtain your cat. Position your cat within the dissection tray, including the tail. Keep any remaining preserving fluid in the bag to keep your cat moist and inhibit bacterial and mold growth.

Procedure

A. Brachial Plexus

1. After placing your cat dorsal side down on the dissecting tray, carefully transect (cut through the middle of) the pectoralis major and minor muscles, if this was not done in the muscle dissection lab.

2. Reflect these muscles to expose the nerves of the brachial plexus. Using a blunt probe, dissect out these nerves.

3. Using Figure C2.1, identify the following four nerves of the brachial plexus: the musculocutaneous, radial, median, and ulnar. Start with the most superior (anterior) nerve in this plexus and work inferiorly (posteriorly).

4. The **musculocutaneous nerve,** the most superior nerve of the brachial plexus, separates into two divisions. The superior division courses under and innervates the coracobrachialis muscle, and the inferior division runs beneath and innervates the biceps brachii muscle.

5. The **radial nerve** is the largest brachial plexus nerve and is located inferior to the musculocutaneous nerve. This nerve innervates the three heads of the triceps muscles, as well as muscles of the forearm.

6. The **median nerve** is inferior to the radial nerve and also follows a similar track as the brachial artery and vein. This nerve continues to innervate muscles of the forearm.

7. The **ulnar nerve** is inferior to the median nerve and continues to innervate muscles of the forearm and front paws.

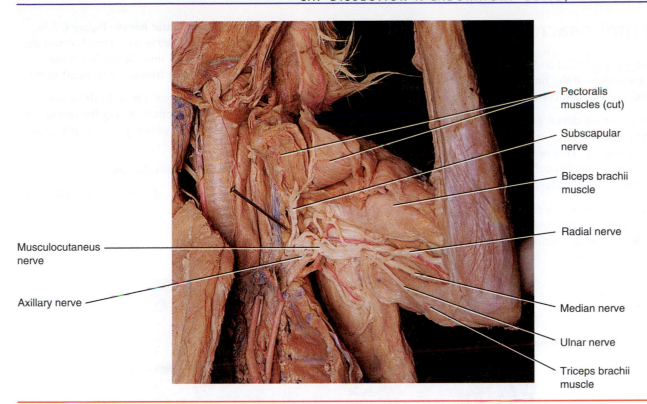

Pectoralis
muscles (cut)

Subscapular
nerve

Biceps brachii
muscle

Radial nerve

Median nerve

Ulnar nerve

Triceps brachii
muscle

Musculocutaneus
nerve

Axillary nerve

FIGURE C2.1 Brachial plexus.

B. Lumbosacral Plexus

1. Using Figure C2.2 and Figure C2.3, identify the four major nerves of the lumbosacral plexus: the femoral, sciatic, tibial, and common peroneal.

2. With your cat dorsal side down, note the **femoral nerve** in the lumbar area emerging from the psoas major muscle. This nerve travels with the femoral artery and vein through the *femoral triangle* and onto the surface of the thigh (Figure C2.2).

3. Turn your cat over with the dorsal side up and transect the biceps femoris muscle, if not done previously in the muscle dissection. Reflect the ends of this mus- cle to expose the wide **sciatic nerve** (Figure C2.3). Follow the course of this nerve as it travels down the posterior thigh and divides into the medial **tibial nerve** and lateral **common fibular (peroneal) nerve.**

4. Place the skin back over your cat and follow your instructor's direction to prepare the cat for storage in the plastic bag. Be sure to attach your group's identi- fication tag.

5. Clean your tabletop with disinfectant.

6. Wash your dissection tools, dissection tray, and hands before leaving the lab.

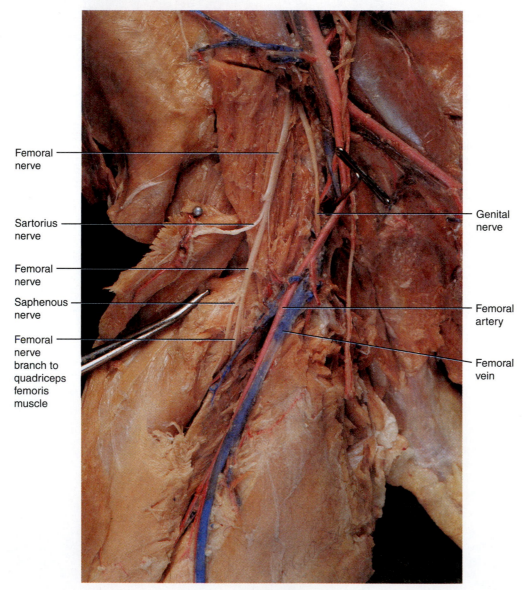

Femoral nerve

Sartorius nerve

Femoral nerve

Saphenous nerve

Femoral nerve branch to quadriceps femoris muscle

Genital nerve

Femoral artery

Femoral vein

Ventral view

FIGURE C2.2 Lumbar plexus.

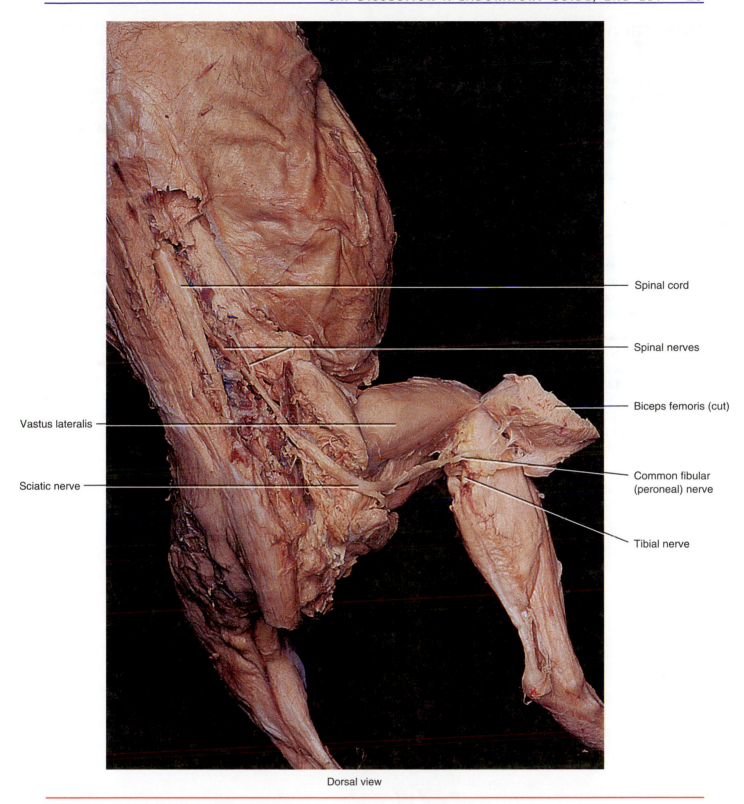

Spinal cord

Spinal nerves

Biceps femoris (cut)

Vastus lateralis

Common fibular
(peroneal) nerve

Sciatic nerve

Tibial nerve

Dorsal view

FIGURE C2.3   Sacral plexus in thigh and leg.

Dissection 3: Endocrine Organs

The major endocrine organs of the cat have similar locations and structure compared with humans. Assemble your dissection equipment and safety glasses, put on your gloves, and obtain your cat. Position your cat within the dissection tray, including the tail. Keep any remaining preserving fluid in the bag to keep your cat moist and to inhibit bacterial and mold growth.

Procedure

1. Place the cat on its back with the ventral side up. Use Figure C3.1 to help you identify the endocrine organs. If you have not opened the ventral body cavities, refer to the instructions in the preface.

2. There are two main endocrine organs in the thoracic cavity: the thyroid and the thymus. The **thyroid gland** has dark lobes similar to the human that are on either side of the trachea inferior to the larynx. The **thymus** has small lobules, and is located inferior to the thyroid gland on the trachea, partially covering the heart. In Figure C3.1, the thyroid gland is dissected to observe one **parathyroid gland** on the dorsal side.

3. There are three main endocrine organs in the abdominal cavity: the pancreas, adrenal glands, and gonads.

Locate the diaphragm that separates the thoracic and abdominopelvic cavities. Reflect the stomach and look beneath it for the light, glandular-looking **pancreas** (Figure C5.2). It is close to the curve in the first part of the small intestine (the duodenum) and extends to the left toward the spleen.

4. The bean-shaped **adrenal glands** are located superior and medial to the kidneys. Both the kidneys and adrenal glands are retroperitoneal, or located behind the peritoneum.

5. The female gonads are called **ovaries** (Figure C8.1c) and are very small, oval organs located inferior to the kidneys.

6. The male gonads, the **testes** (Figures C8.1a and C8.1b), are located outside of the abdominopelvic cavity in the scrotum. Before opening the scrotum, your instructor will tell you whether or not to proceed to view the testes at this time, or to wait for the reproductive system.

7. Place the skin back over your cat and follow your instructor's directions to prepare the cat for storage in the plastic bag. Be sure to attach your group's identification tag.

8. Clean your tabletop with disinfectant.

9. Wash your dissection tools, dissection tray, and hands before leaving the lab.

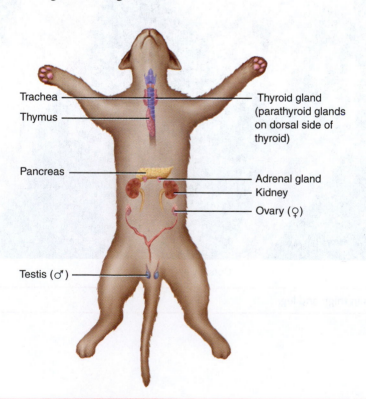

Trachea

Thymus

Pancreas

Testis (♂)

Thyroid gland (parathyroid glands on dorsal side of thyroid)

Adrenal gland

Kidney

Ovary (♀)

FIGURE C3.1 Endocrine glands.

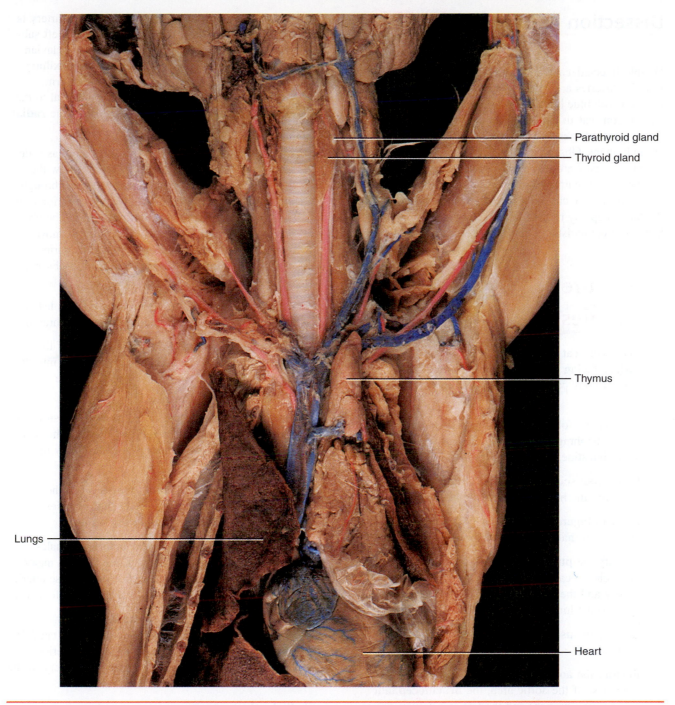

FIGURE C3.1 Endocrine glands, *continued*.

Dissection 4: Blood Vessels

Double-injected cats are usually used to identify blood vessels. Arteries are injected with red latex, and veins are injected with blue latex. Blood vessels differ slightly in location from cat to cat. It is important to understand that these slight differences in location are normal and also occur in humans. Observe the fascia that protects and secures blood vessels. Carefully remove the fascia with blunt instruments to separate blood vessels from other structures.

Position your cat within the dissection tray, including the tail. Keep any remaining preserving fluid in the bag to keep your cat moist and inhibit bacterial and mold growth.

Procedure

A. Arteries

1. Place your cat in a dissecting tray with the ventral surface facing upward. If you have not opened the ventral body cavities, refer to the instructions in the preface.

2. Identify the following major organs: heart, trachea, lungs, diaphragm, stomach, spleen, pancreas, liver, small intestine, and large intestine.

3. Using your scissors, cut open the pericardial sac surrounding the heart to expose the heart.

4. Refer to Figure C4.1 to identify the arteries listed in steps 5–10 that are located above the diaphragm.

5. Identify the **pulmonary trunk** exiting from the right ventricle. Locate its branches, the **right pulmonary artery** and the **left pulmonary artery,** and follow them to the lungs.

6. Identify the **ascending aorta** as it exits the left ventricle.

7. Identify the **aortic arch.** In cats, there are only two branches off the aortic arch, the **brachiocephalic artery** (first branch) and the **left subclavian artery.** Identify these branches. Compare this branching with the human.

8. The brachiocephalic artery divides into the right subclavian artery, the right common carotid, and the left common carotid. Locate the **subclavian artery** as it turns laterally and travels toward the upper extremity. Locate the **right** and **left common carotid arteries** as they travel along the trachea. At the level of the larynx, the common carotid arteries divide to form the external and internal carotid arteries.

9. The first major branch off each subclavian artery is the **vertebral artery.** Follow the right and left subclavian arteries to the first rib. As each subclavian artery crosses the first rib, it becomes the **axillary artery.** Follow the axillary artery into the arm, where it becomes the **brachial artery.** Distal to the elbow, the brachial artery divides to form the **radial** and **ulnar arteries.**

10. Lift up the heart and follow the aortic arch as it descends and forms the **thoracic aorta.** Follow the thoracic aorta and observe where it passes through the diaphragm with the esophagus and inferior vena cava, and becomes the **abdominal aorta.** The abdominal aorta is retroperitoneal. You must move aside the visceral organs and remove the parietal peritoneum lining the dorsal body wall to observe the aorta.

11. Refer to Figure C4.2 to identify the arteries listed in steps 12–20 that are located below the diaphragm.

12. Locate the **celiac trunk,** the first branch off the abdominal aorta. The celiac trunk branches into the hepatic artery, the left gastric artery, and the splenic artery.

13. Posterior (caudal) to the celiac trunk is the **superior (anterior) mesenteric artery,** whose branches can be observed traveling through the mesentery of the small intestine.

14. Follow the abdominal aorta to the level of the kidneys and observe the paired **renal arteries** branching off and traveling to the kidneys.

15. The gonadal arteries, **testicular arteries** in males and **ovarian arteries** in females, are the next major branches off the abdominal aorta. Follow these arteries to the gonads (testes in males and ovaries in females).

16. The **inferior (posterior) mesenteric artery** branches off the abdominal aorta posterior (caudal) to the gonadal arteries. Branches of the inferior mesenteric artery travel through the mesentery of the large intestine.

17. **Iliolumbar arteries** are large branches off the abdominal aorta posterior to the inferior mesenteric arteries.

18. The abdominal aorta ends when it divides into the **right** and **left external iliac arteries,** and the **internal iliac artery.** There is no common iliac artery in the cat.

19. Follow one external iliac artery into a thigh, where it becomes the **femoral artery.**

20. The femoral artery travels down the thigh and becomes the **popliteal artery** in the popliteal area.

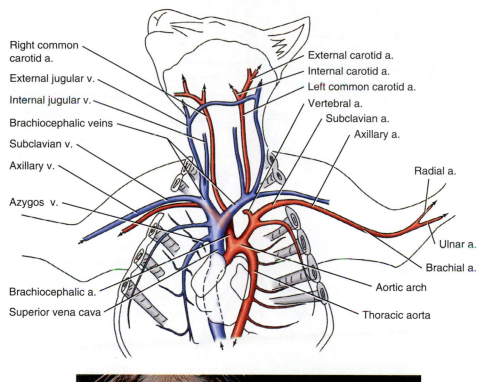

Right common carotid a.
External jugular v.
Internal jugular v.
Brachiocephalic veins
Subclavian v.
Axillary v.
Azygos v.
Brachiocephalic a.
Superior vena cava

External carotid a.
Internal carotid a.
Left common carotid a.
Vertebral a.
Subclavian a.
Axillary a.
Radial a.
Ulnar a.
Brachial a.
Aortic arch
Thoracic aorta

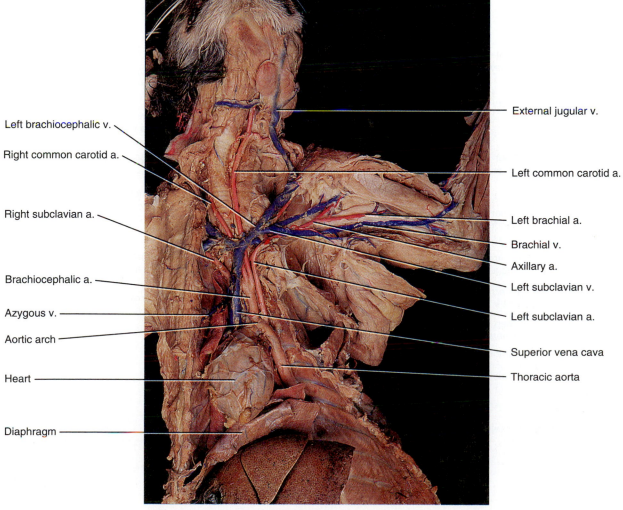

Left brachiocephalic v.
Right common carotid a.
Right subclavian a.
Brachiocephalic a.
Azygous v.
Aortic arch
Heart
Diaphragm

External jugular v.
Left common carotid a.
Left brachial a.
Brachial v.
Axillary a.
Left subclavian v.
Left subclavian a.
Superior vena cava
Thoracic aorta

Ventral view

FIGURE C4.1   Blood vessels above the diaphragm.

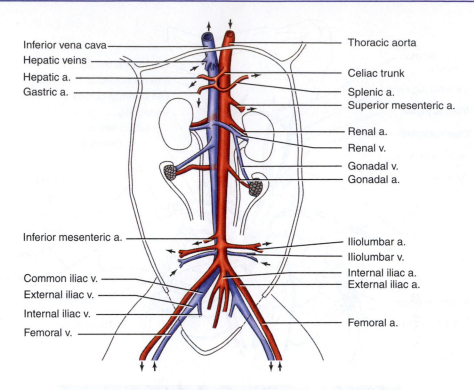

Inferior vena cava — — Thoracic aorta
Hepatic veins — — Celiac trunk
Hepatic a. — — Splenic a.
Gastric a. — — Superior mesenteric a.
— Renal a.
— Renal v.
— Gonadal v.
— Gonadal a.
Inferior mesenteric a. — — Iliolumbar a.
— Iliolumbar v.
— Internal iliac a.
Common iliac v. — — External iliac a.
External iliac v. —
Internal iliac v. —
Femoral v. — — Femoral a.

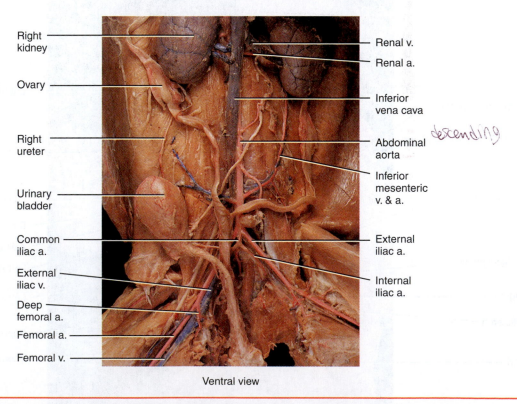

Right kidney — — Renal v.
— Renal a.
Ovary — — Inferior vena cava
Right ureter — *descending* — Abdominal aorta
— Inferior mesenteric v. & a.
Urinary bladder —
Common iliac a. — — External iliac a.
External iliac v. — — Internal iliac a.
Deep femoral a. —
Femoral a. —
Femoral v. —

Ventral view

FIGURE C4.2   Blood vessels below the diaphragm.

B. Veins

1. Blood leaving tissues travels through veins back to the heart. Remember that some veins are superficial (close to the surface), whereas others are deep. Many of the deep veins are adjacent to arteries with the same name.

2. Refer to Figure C4.2 to identify veins located caudal to the diaphragm.

3. Observe the large superficial vein traveling along the medial surface of the leg ascending into the thigh. This is the **great saphenous vein,** and it joins the **femoral vein,** a deep vein, traveling through the thigh adjacent to the femoral artery.

4. The femoral vein becomes the **external iliac vein** in the groin region. The internal iliac vein joins the external iliac vein to form the **common iliac vein.**

5. The right and left common iliac veins unite to form the **inferior vena cava (postcava** in cat).

6. The **renal veins** and **gonadal veins** carry blood from the kidneys and gonads to the inferior vena cava.

7. The **hepatic portal vein** probably does not contain blue latex and may appear brown from the presence of coagulated blood. The hepatic portal vein receives blood from the digestive organs and carries this blood to the liver. The hepatic portal vein is formed from the gastrosplenic vein and the superior mesenteric vein.

8. Follow the inferior vena cava through the diaphragm, into the thoracic cavity, and into the right atrium.

9. Refer to Figure C4.1 to identify veins cephalic to the diaphragm.

10. Locate the **radial** and **ulnar** veins in the forearm. These veins are adjacent to their corresponding arteries. The radial and ulnar veins merge to form the **brachial vein.**

11. The brachial vein becomes the **axillary vein** that is adjacent to the axillary artery in the axillary regions.

12. In the shoulder area, the axillary vein becomes the **subclavian vein.**

13. Each subclavian vein unites with an external jugular vein to form either the **right** or **left brachiocephalic vein.** The brachiocephalic veins merge to form the **superior vena cava (precava).** Follow the superior vena cava unit it enters the right atrium.

14. Blood draining from the face and skull enters the external jugular vein. The internal jugular vein drains the brain. Identify the large **external jugular vein** traveling along the lateral surface of the neck until it joins with the subclavian vein to form the brachiocephalic vein.

15. Place the skin back over your cat and follow your instructor's directions to prepare the cat for storage in the plastic bag. Be sure to attach your group's identification tag.

16. Clean your tabletop with disinfectant.

17. Wash your dissection tools, dissection tray, and hands before leaving the lab.

Dissection 5: Lymphatic System

The lymphatic system of the cat is complementary to the human with the organs being similar in location and structure compared with the human. Assemble your dissection equipment and safety glasses, put on your gloves, and obtain your cat. Position your cat within the dissection tray, including the tail. Keep any remaining preserving fluid in the bag to keep your cat moist and inhibit bacterial and mold growth.

Procedure

1. You may have already looked at the lymphatic organs in your previous dissections. If your cat is triple injected with yellow or green latex for the lymphatic system, it will be easier to find the lymphatic organs and very thin vessels. Use Figure C5.1 to identify lymphatic glands and organs.

2. As you dissected the blood vessels, you may have noted small, bean-shaped **lymph nodes** in the cervical, axillary, and inguinal areas. Because these nodes are small, they are easy to miss if you do not know their structure or location.

3. The noncapsulated **thymus** is over the anterior surface of the heart and sometimes is also a little superior to the heart. This gland may have been identified in the endocrine system.

4. The **spleen** is located in the upper left quadrant posterior and lateral to the stomach, and may be a dark brownish-red color.

5. The **thoracic duct** (left lymphatic duct) can sometimes be found in the dorsal part of the thoracic cavity, especially if your cat has been triple injected with latex. Move the lungs and heart aside and look just to the left of the midline next to the descending (thoracic) aorta. The thoracic duct will be very thin (1/16 inch) and may be reddish-brown, with a segmented look that is caused by the presence of valves. You may be able to trace it to where it enters the junction of the left subclavian and external jugular veins. The right lymphatic duct is smaller and is not as easy to find.

6. Place the skin back over your cat and follow your instructor's directions to prepare your cat for storage in the plastic bag. Be sure to attach your group's identification tag to the cat or bag.

7. Clean your laboratory tabletop with disinfectant.

8. Wash your dissection tools, dissection tray, and hands before leaving the lab.

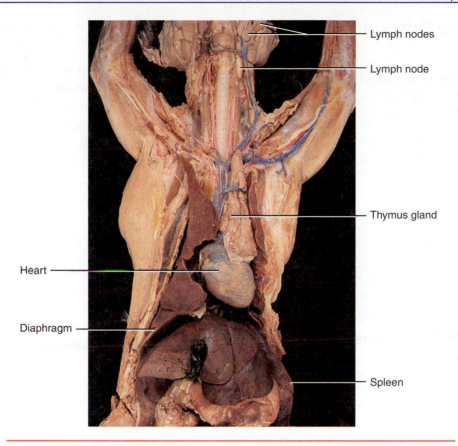

FIGURE C5.1 Lymphatic glands and organs.

Dissection 6: Respiratory System

The respiratory system of the cat is complementary to the human. The structure of the larynx, trachea, lungs, and diaphragm are similar to the human. Assemble your dissection equipment and safety glasses, put on your gloves, and obtain your cat. Make sure all parts of the cat are inside the dissection tray, including the tail. Keep any remaining preserving fluid in the bag to keep your cat moist and inhibit bacterial and mold growth.

Procedure

1. Use Figure C6.1 to help you identify the bolded structures listed below in the cat.

2. Observe the **external nares** (choanae), **nasal cavity,** and **oral pharynx.**

3. Locate the **larynx,** the prominent **thyroid cartilage** in the anterior neck region, and the **cricoid cartilage** inferior to the thyroid cartilage. Use the blunt probe to separate the larynx from the muscles and connective tissue.

4. Your instructor may divide the lab groups in half to observe two different views of the larynx as listed below:
   - Half of the lab groups will cut the complete larynx away from the laryngopharynx at the hyoid bone. Pull the larynx toward you, look into the top of the larynx, and identify: the **epiglottis** (elastic cartilage), **glottis, false vocal cords** (anteriorly), and **true vocal cords** (posteriorly).
   - The other half of the lab groups will make a longitudinal cut through the thyroid cartilage, the larynx, and through the superior part of the trachea. Observe the following structures: the **epiglottis, glottis, false vocal cords,** and **true vocal cords.**

5. Examine the **trachea,** following it into the thoracic cavity. Feel the **C-shaped tracheal cartilages.** Check to see if the thyroid gland is still present, or if it was removed in a previous dissection.

6. Cut the trachea in cross section and pull the cut portion toward you. Carefully separate the connective tissue between the esophagus and the trachea with a blunt probe. Observe the dorsal side of the trachea and identify the **trachealis muscle** that connects the free edges of the tracheal cartilages.

7. If you have already studied the cardiovascular system, ask your instructor for permission to remove the heart and great vessels from the cat.

8. With the heart removed, you can easily identify the end of the trachea in the thoracic cavity at its bifurcation into the **right** and **left primary bronchi.**

9. Dissect away lung tissue on the left side to follow the left primary bronchus as it branches into the **secondary bronchi.** If you keep dissecting, you may want to use a dissecting microscope to observe smaller **tertiary bronchi.**

10. On the right side, you should find the **anterior, medial, posterior,** and **mediastinal lobes** of the lung. The latter lobe will be more midline.

11. On the left side, you should find the **anterior, medial,** and **posterior lobes.**

12. Identify the **hilus** of the lung on its medial border, along with the **primary bronchus, pulmonary artery,** and **pulmonary veins.**

13. Look deep into the thoracic cavity and identify the shiny **parietal pleura** that covers the ribs and intercostal muscles. The **visceral pleura** also glistens and covers the lungs themselves.

14. Observe the muscular **diaphragm** that forms the thoracic cavity floor and its relationship to the lungs and the pleura of the lungs.

15. Place the skin back over your cat and follow your instructor's directions to prepare your cat for storage in the plastic bag. Be sure to attach your group's identification tag to the cat or bag.

16. Clean your tabletop with disinfectant.

17. Wash your dissection tools, dissection tray, and hands before leaving the lab.

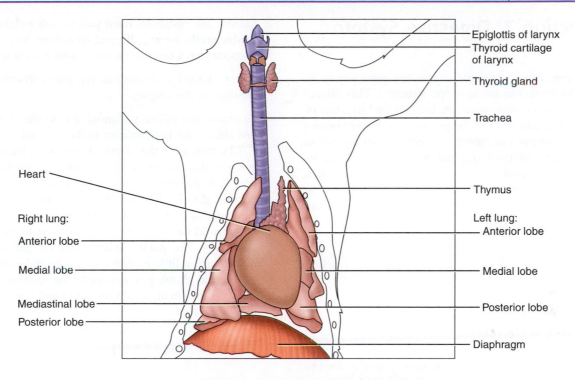

Epiglottis of larynx
Thyroid cartilage of larynx
Thyroid gland
Trachea
Thymus

Heart

Right lung:

Anterior lobe

Medial lobe

Mediastinal lobe

Posterior lobe

Left lung:
Anterior lobe

Medial lobe

Posterior lobe

Diaphragm

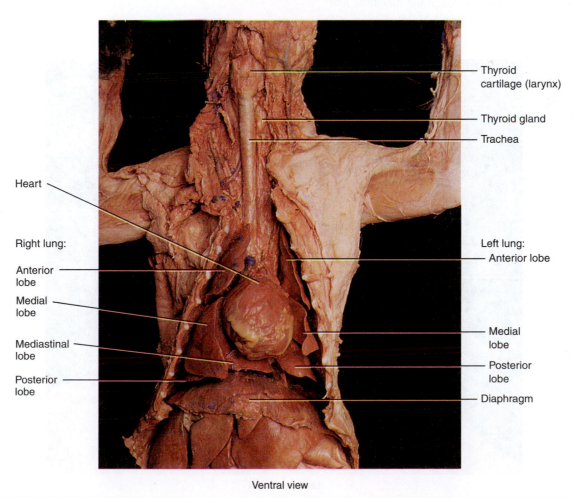

Thyroid cartilage (larynx)

Thyroid gland

Trachea

Heart

Right lung:

Anterior lobe

Medial lobe

Mediastinal lobe

Posterior lobe

Left lung:
Anterior lobe

Medial lobe

Posterior lobe

Diaphragm

Ventral view

FIGURE C6.1   Respiratory system, ventral view.

Dissection 7: Digestive System

The cat digestive system and organs are quite similar to that of the human in location and structure. This dissection also clearly demonstrates the location and structure of the mesentery and parts of the peritoneum that are not realistically portrayed in models. Assemble your dissection equipment and safety glasses, put on your gloves, and obtain your cat. Position your cat within the dissection tray, including the tail. Keep any remaining preserving fluid in the bag, to keep your cat moist and to inhibit bacterial and mold growth.

Procedure

A. Mouth, Oropharynx, and Salivary Glands

1. To observe the oral cavity structures, you may need to use a bone cutter to cut through the mandible and separate it from the maxilla.

2. Identify the **vestibule, hard palate, soft palate, canine teeth, tongue, lingual frenulum,** and **oropharynx.** Compare the teeth with human teeth.

3. Using a hand lens or magnifying glass, observe the papillae on the tongue.

4. To expose the **salivary glands,** remove the skin on one side of the head inferior to the ear (see Figure C.7.1), trim away the connective tissue in the area between this and the masseter muscle. Look for tiny, dark **lymph nodes** (bean-shaped) in this area. The **parotid gland** is a light-colored gland on the cheek area inferior to the ear. You may be able to identify the **parotid duct** traversing the masseter to enter the oral cavity. The smaller **submandibular gland** is inferior and a little posterior to the parotid gland. The **sublingual gland,** just anterior to the submandibular gland, is the smallest salivary gland and is more difficult to find.

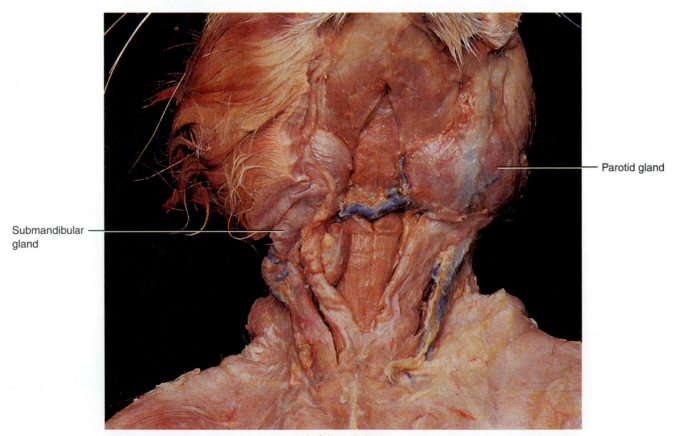

Submandibular gland

Parotid gland

Left lateral view

FIGURE C7.1   Salivary glands.

B. Esophagus and Abdominal Organs

1. If you have dissected the respiratory system, you have previously observed the laryngopharynx, epiglottis, larynx, and trachea. The **laryngopharynx** also leads to the **esophagus** that is posterior to the trachea. Follow the esophagus through the thoracic cavity to the **diaphragm,** locating the **esophageal hiatus** where the esophagus penetrates through the diaphragm to the abdominal cavity.

2. Use Figure C.7.2a and Figure C.7.2b as a reference to identify the bolded structures.

3. Observe the yellowish, fat-filled "apron" that covers the abdominopelvic viscera. This is the double-layered serous membrane, the **greater omentum,** that can be deflected back or totally removed according to your instructor's directions.

4. Observe the **peritoneum** that lines the abdominal cavity and also covers the exterior of the abdominal organs. The **peritoneal cavity** is the large cavity that is filled with the abdominopelvic organs.

5. The next obvious structure in the abdomen is the large, brown or reddish-brown **liver** on the right side inferior to the diaphragm. Look for a small, greenish sac, the **gallbladder,** on the inferior surface of the liver, and the **cystic duct.** The **falciform ligament** separates the right and left lobes of the liver and attaches the liver superiorly to the abdominal wall.

6. To the left of and partially posterior to the liver is the **stomach.** Identify the **lesser omentum,** the serous membrane that attaches the liver to the **lesser curvature** of the stomach. Note the constricted junction of the esophagus and the stomach, the **esophageal sphincter.** Cut open the stomach along its **greater curvature** to reveal the **rugae,** if present. If the cat's stomach is stretched, rugae are absent; if the stomach is contracted, rugae will be present. Identify the parts of the stomach: the **cardia, fundus, body, pylorus,** and the **pyloric sphincter.** Roll the firm sphincter area between your thumb and

index finger; cut open this area to observe the constriction caused by the sphincter. To the left of and posterior to the stomach is the long, narrow, dark-colored **spleen** that hugs the left abdominal wall (not a digestive organ).

7. Lift the stomach, and reflect it back to reveal the granular, usually brownish-gray **pancreas.** The **head** of the pancreas is in the C-shape of the first section of the small intestine, the **duodenum,** and the **tail** of the pancreas is near the spleen. Find the **common bile duct** entering the duodenum and follow it toward the liver until you find the junction of the **common hepatic duct** with the cystic duct.

8. The small intestine of the cat has three divisions, as does the human: the **duodenum, jejunum,** and **ileum.** Note the **mesentery** that attaches the small intestine to the posterior body wall. Spread the mesentery to observe the branches of the superior mesenteric artery and vein. Follow the small intestine through its entire length. The ileum ends in the inferior right quadrant, where it joins with the large intestine at the **ileocecal junction** or **sphincter.** Make an incision in this area to observe the sphincter. Note that the small intestine has a smaller diameter, a greater length, and is very coiled compared with the large intestine.

9. The **large intestine,** or **colon,** is composed of the **cecum,** a short **ascending colon, transverse colon, descending colon,** and **rectum.** Just inferior to the ileocecal junction is the **cecum,** or blind pouch. Identify the **ascending, transverse,** and **descending** parts of the **colon** plus the **mesocolon** that affixes the colon to the posterior body wall. Now identify the **rectum** and the **anus.**

10. Place the skin back over your cat and follow your instructor's directions to prepare your cat for storage in the plastic bag. Be sure to attach your group's identification tag to the cat or bag.

11. Clean your laboratory tabletop with disinfectant.

12. Wash your dissection tools, dissection tray, and hands before leaving the lab.

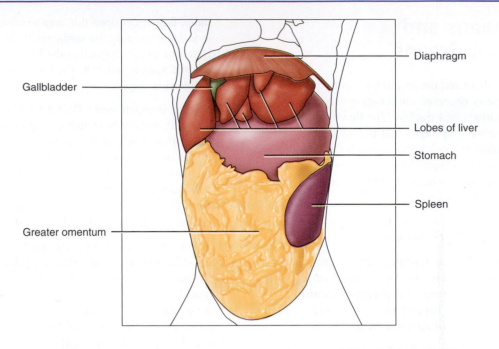

Gallbladder

Diaphragm

Lobes of liver

Stomach

Spleen

Greater omentum

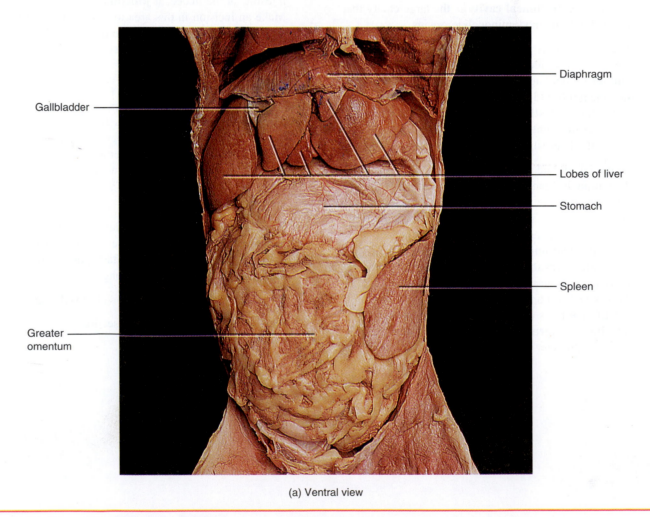

Gallbladder

Diaphragm

Lobes of liver

Stomach

Spleen

Greater omentum

(a) Ventral view

FIGURE C7.2a Digestive organs, superficial.

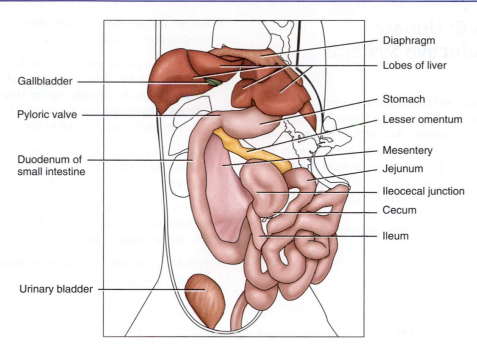

Gallbladder

Pyloric valve

Duodenum of
small intestine

Urinary bladder

Diaphragm

Lobes of liver

Stomach

Lesser omentum

Mesentery

Jejunum

Ileocecal junction

Cecum

Ileum

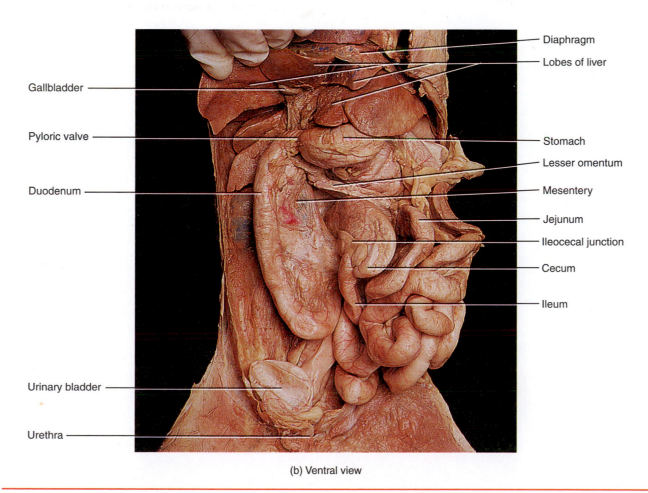

Gallbladder

Pyloric valve

Duodenum

Urinary bladder

Urethra

Diaphragm

Lobes of liver

Stomach

Lesser omentum

Mesentery

Jejunum

Ileocecal junction

Cecum

Ileum

(b) Ventral view

FIGURE C7.2b Digestive organs, deep.

Dissection 8: Urinary and Reproductive Systems

Typically, the urinary and reproductive systems are studied together, because of their close association of structures and their embryologic derivations. The urinary and reproductive systems of the male cat are similar to the human. The female cat has more differences compared to the human, because she has litters rather than one offspring during one pregnancy. Assemble your dissection equipment and safety glasses, put on your gloves, and obtain your cat. Position your cat within the dissection tray, including the tail.

Procedure

A. Urinary System

1. Refer to Figure C8.1a if you have a male cat, or Figure C8.1b if you have a female. Identify the **bolded** urinary structures described.

2. Reflect the abdominal viscera that were observed in the digestive system dissection.

3. Remove the peritoneum from the kidneys if not removed in a prior dissection and carefully remove the **adipose capsule** surrounding the kidneys. Locate the **adrenal glands** that are not attached to the kidneys, but are superior and medial to them.

4. Locate the **renal hilus** on the medial surface of each kidney and identify the **renal artery, renal vein,** and the **ureter** passing through the renal hilus.

5. Follow the renal artery to where it branches off the **abdominal aorta** and the renal vein to where it enters the inferior **vena cava.**

6. Follow the ureters to the **urinary bladder,** a retroperitoneal, muscular sac. If you have a female cat, be careful not to mistake the uterine horns for the ureters. Observe the entrance of the **ureters** into the posterior wall of the urinary bladder, and the peritoneal folds that secure the urinary bladder to the abdominal wall.

7. The **urethra** will not be dissected out at this time because of its location, but will be located in the reproductive system dissection that follows.

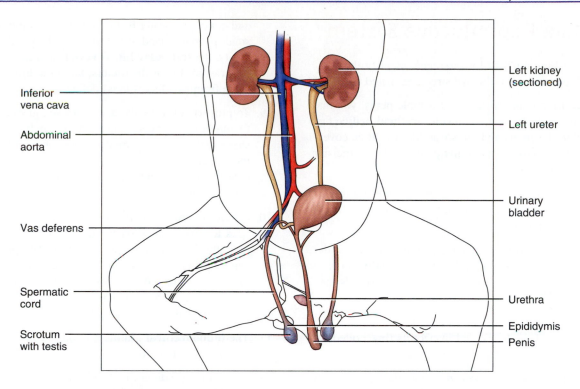

Inferior vena cava

Abdominal aorta

Vas deferens

Spermatic cord

Scrotum with testis

Left kidney (sectioned)

Left ureter

Urinary bladder

Urethra

Epididymis

Penis

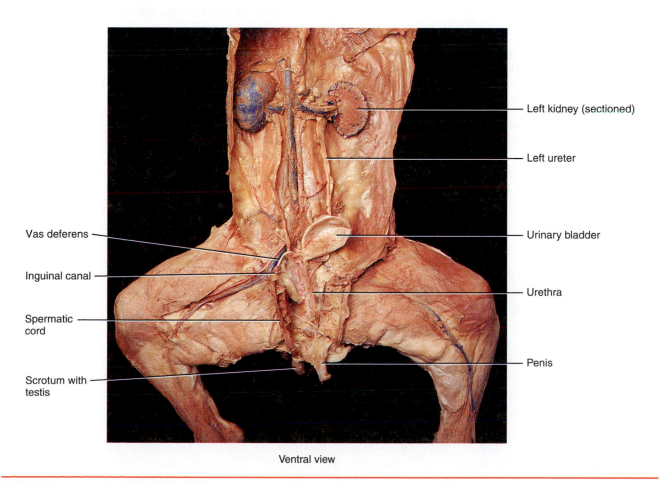

Vas deferens

Inguinal canal

Spermatic cord

Scrotum with testis

Left kidney (sectioned)

Left ureter

Urinary bladder

Urethra

Penis

Ventral view

FIGURE C8.1a Male urogenital system.

B. Male Reproductive System

1. Using Figure C8.1a for reference, identify the **bolded** male reproductive structures listed below.

2. Because a male cat has a retractable penis, you may need to check for the **external urethral orifice** first to find the **penis** and the sheathlike **prepuce** covering it. To observe the **glans penis,** make an incision in the prepuce.

3. Identify the **scrotum** or scrotal sac covering the paired **testes,** which may not be very obvious if you have a young male.

4. Carefully, make a lateral incision in one side of the scrotum and remove the loose fascia and inner fibrous connective tissue to expose one testis. Is the scrotal sac open to both testes?

5. Note the **epididymis** on the medial and posterior surfaces of the testis, and inspect its tiny, coiled tubules. You may want to use a hand lens for this.

6. Identify the **ductus (vas) deferens** that begins at the tail of the epididymis and travels toward the body in the spermatic cord.

7. Observe the **spermatic cord** and cut away the connective tissue to identify the **ductus (vas) deferens, testicular artery, testicular vein,** and **autonomic nerves** within it. Follow the ductus (vas) deferens through the **inguinal canal** into the pelvic cavity.

8. Trace the path of the ductus (vas) deferens in the abdominopelvic cavity as it arches around the ureter,

and continues posterior to the bladder to join the small prostate gland at the **urethra.** Inside the pelvic cavity, the **testicular blood vessels** and **autonomic nerves** travel near the ureters, taking a different route from the ductus (vas) deferens.

9. To properly observe the accessory sex glands and the urethra, you need to cut the cat's pelvis. Using a sharp scalpel, make a midline incision to cut through the muscles covering the symphysis pubis and then carefully cut through the center of the pubic symphysis cartilage.

10. Spread the thighs apart and bend the pelvic bones back to expose the prostate gland, paired bulbourethral glands, urethra, and penis. The **prostate** can be palpated as a small, hard mass surrounding the urethra. The cat anatomy is similar to, but not identical to, the human. There are no seminal vesicles in the cat.

11. The **bulbourethral glands** are located posterior to the prostate gland dorsal to the penis.

12. Make a longitudinal incision in the penis and identify the two columns of **corpora cavernosa,** one column of **corpus spongiosum,** and the **spongy urethra.**

13. Observe the dissection of a female cat from another lab group. Read steps 9–11 of the female cat reproductive system dissection for clean-up directions.

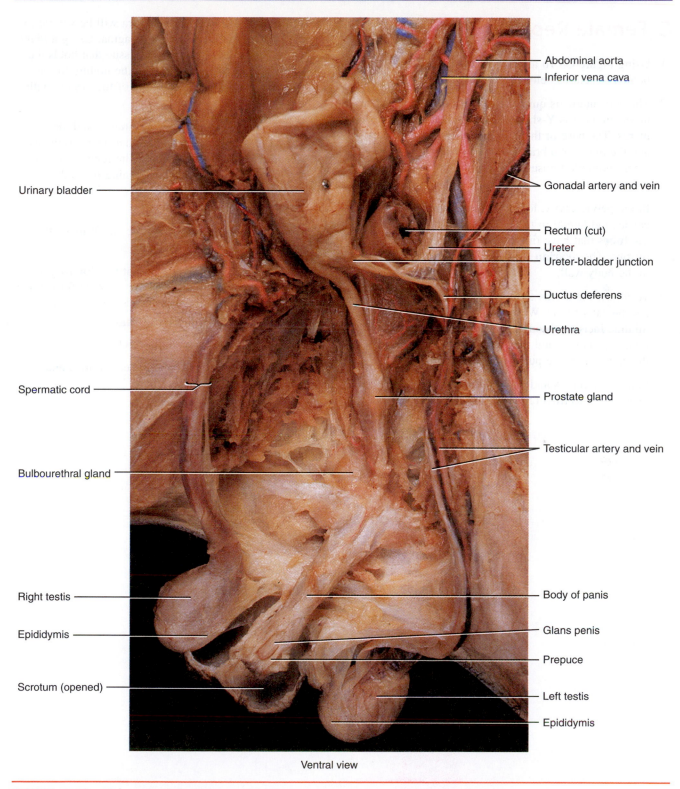

Abdominal aorta

Inferior vena cava

Urinary bladder

Gonadal artery and vein

Rectum (cut)

Ureter

Ureter-bladder junction

Ductus deferens

Urethra

Spermatic cord

Prostate gland

Testicular artery and vein

Bulbourethral gland

Right testis

Body of panis

Epididymis

Glans penis

Prepuce

Scrotum (opened)

Left testis

Epididymis

Ventral view

FIGURE C8.1b Male urogenital system, *continued.*

C. Female Reproductive System

1. Using Figure C8.1b as a reference, identify the bolded structures below.

2. The cat's uterus is quite different from a human. The uterus in a cat is Y-shaped and is called a bipartate uterus. The base of the Y is the **body of the uterus** and the upper two branches are the **uterine horns** where multiple fetuses may be located if your cat is pregnant.

3. In the pelvic cavity, locate the small, oval **ovaries** caudal and lateral to the kidneys and the small **uterine tubes** that have tiny **fimbriae** curved over the ovaries. The **ovarian ligament** attaches the ovaries to the body wall.

4. To follow the uterus to the vagina, you will need to cut the cat's pelvis. With a sharp scalpel, make a midline incision through the muscles covering the pubic symphysis and then cut through the center of the cartilage of the pubic symphysis.

5. Spread the thighs and bend the pelvic bones back to expose the **urethra** (anterior) and **vagina** (posterior).

6. The **urinary bladder** and urethra will be ventral to the body of the uterus and the vagina. Using a blunt probe, separate the connective tissue that holds the urethra to the vagina and move the urethra to the side. Locate the posterior union of the urethra with the vagina.

7. Just caudal to the union of the urethra and the vagina is the **urogenital sinus** that opens to the exterior in the **urogenital oriface.** The female cat has the urogenital orifice as one opening for both the urinary and reproductive systems similar to the male cat and human male.

8. Observe the dissection of a male cat from another lab group.

9. Place the skin back over your cat and follow your instructor's directions to prepare your cat for storage in the plastic bag. Be sure to attach your group's identification tag to the cat or bag.

10. Clean your tabletop with disinfectant.

11. Wash your dissection tools, dissection tray, and hands before leaving the lab.

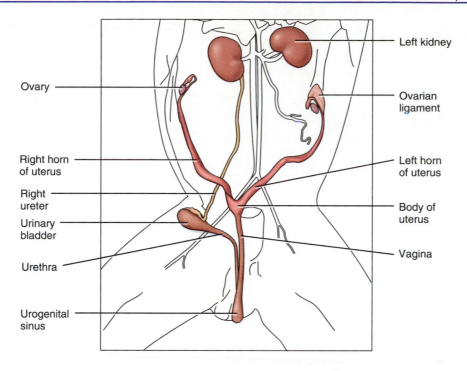

Ovary

Right horn
of uterus

Right
ureter

Urinary
bladder

Urethra

Urogenital
sinus

Left kidney

Ovarian
ligament

Left horn
of uterus

Body of
uterus

Vagina

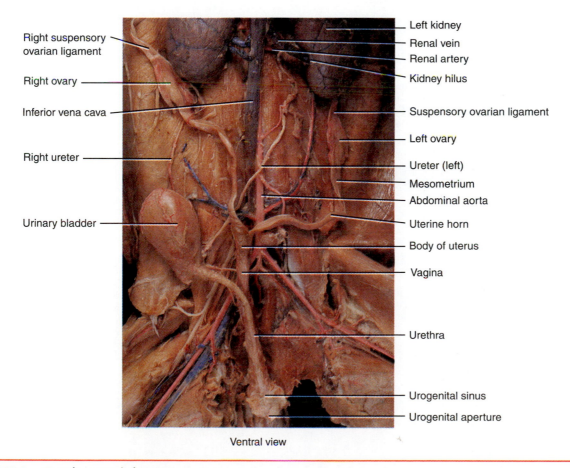

Right suspensory
ovarian ligament

Right ovary

Inferior vena cava

Right ureter

Urinary bladder

Left kidney
Renal vein
Renal artery
Kidney hilus

Suspensory ovarian ligament

Left ovary

Ureter (left)
Mesometrium
Abdominal aorta
Uterine horn

Body of uterus

Vagina

Urethra

Urogenital sinus
Urogenital aperture

Ventral view

FIGURE C8.1c   Female urogenital system.